Here are some things to do at Kew...

Feel what it is like to be in a steamy jungle in the Palm House.

Discover plants that eat animals in the Princess of Wales Conservatory.

Climb, scramble and swing amongst the Treehouse Towers – suitable for 3-11 year olds. Play in Climbers and Creepers – if you are 9 years or younger.

Take a wildlife walk in the Conservation Area.

Be a badger in our human-sized Badger Sett.

Climb up the Xstrata Treetop Walkway – if you are not afraid of heights! Go down into the Rhizotron to find out about tree roots and life underground.

Why does Kew have so many plants?
We grow some to keep them safe because they are rare or even extinct in the wild. Others are grown to show how important plants are to people and the rest of life on the planet.

Palm House

Inside the Palm House, it is like
the steamy tropical rainforest.

Did you know?
The Palm House has 16,000 panes
of glass. No wonder it takes
weeks to clean them all!

Babassu palm
the largest palm in
the Palm House

Panama hat plant
see Plants and
People Exhibition

Bottle palm
see Plants in danger

Coco-de-mer

Annatto tree

**Cacao
tree**

Coffee bush

**Coconut
palm**

**Vanilla
orchid**

Café marron
see Plants in danger

Cycad
Kew's oldest pot plant

Giant bamboo

Put a sticker next to each plant on the list that you spot on your jungle jaunt.

Find the plants

Mango tree
only produces fruit in the tropics

Sugar cane

Rattan palm

Marine
Display

Banana
look for green
unripe bananas

Rubber
tree

 Annatto tree

 Cacao tree

 Coconut palm

 Coffee bush

 Cycad

 Giant bamboo

 Rattan palm

 Rubber tree

 Coco-de-mer

 Sugar cane

 Vanilla orchid

The lake is this way.

You can find out about these plants on the next few pages.

3

Inside the Palm House

Find rainforest plants that are ingredients for some of the world's favourite foods and drinks.

Chocolate, yum!

Cocoa beans grow inside pods that sprout from the trunk of the cacao tree. The beans are piled into warm, soggy heaps to bring out their delicious chocolate flavour. They are dried and shipped to chocolate factories.

Cacao beans

Cacao tree

Coconut mat

Curries, cakes and doormats

Coconuts are the seeds of the coconut palm. The white kernel inside is used in many foods from curries to cakes. The coarse fibres on the outside of the coconut can be made into doormats.

Coconut

Curry

Coconut palm

CHEESE

Annatto or bixa tree.

Food colouring

The orange colour in many foods comes from the soft flesh that surrounds the seeds of the annatto tree. In tropical America, native people use the colour as body paint and lipstick.

Did you know?
A coconut can kill, if it falls on a person's head!

Ice-cream flavour

Vanilla flavour comes from the seed pods of this tropical climbing orchid. The pods have no flavour until they are sweated and dried. Natural vanilla is expensive so most vanilla flavouring is artificial.

Vanilla orchid

How many ants can you see on this page?
Answer at the back of the guide.

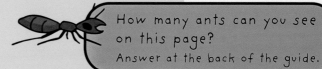

Wakey, wakey!

Most of the world's coffee comes from bushes grown in tropical America. There are two coffee beans inside each red fruit. The beans need to be roasted before they smell and taste of coffee.

Coffee bush

Sweet grass

Sugar cane is a tall grass, which grows in the tropics. The canes are shredded and crushed to release the sweet juice inside. Over two-thirds of the world's sugar is made from sugar cane.

Fizzy drinks can have loads of sugar.

Sugar cane

What do you call shoes made of banana skins?
Slippers

Turn to page 44 to find out about some of the rare rainforest plants in the Palm House.

Inside the Palm House

Look for these rainforest plants, which help us to build and make things. You can also find some record-breakers here.

Cutting rattan

Rattan

Rubber balls

A tapper collecting latex

Cane chairs

Rattans are climbing palms with long stems that are made into cane chairs, baskets and mats. The palm has spines that help it to scramble high in the forest canopy. The spines may hurt collectors harvesting rattans from the rainforest.

Rubber tree

Bouncy balls

White liquid, called **latex**, oozes from cuts made in the bark of the rubber tree. The latex is dried and made into stretchy rubber. Balls made of rubber bounce because they spring back into shape.

Coco-de-mer

Coco-de-mer seed

Big bum

The Coco-de-mer palm has the world's heaviest seed, which looks like a big bottom. The seed weighs up to 30 kilos. After a seed thuds to the ground, it takes over two years before it starts to grow.

Triceratops

Giant bamboo

Really ancient

Cycads are a group of plants that thrived 200 million years ago, when dinosaurs liked to eat them! This cycad in the Palm House is over 200 years old.

The cycad is Kew's oldest pot plant.

Fast grass

Giant bamboos are the world's fastest growing plants. They grow up to 25 metres high, which is as tall as a stack of 16 cars. Bamboo is used for scaffolding in south-east Asia because it is strong and flexible.

Bamboo scaffolding

I'm a rainforest pitcher plant. Unscramble the words to find plants you might need if you got lost in the rainforest without supplies.

Clue — all the plants are on the Palm House map!

1. Make a shelter using **alpm** leaves and **omobab** poles.

2. Take cane from the **tartan** palm to make a chair.

3. Collect some **tonococus**, **sanbnaa**, **asrgu neca**, and **esnoamg** to eat.

Answers at the back of the guide

Palm House Marine Display

Head downstairs in the Palm House to discover plants that live on the shore and in the sea.

Coral reef

Some tiny plants are partners with animals, called corals. They help the corals build huge reefs in tropical seas. Coral reefs are home to many kinds of fishes and other animals. Most reef fishes are brightly coloured to signal to others on the crowded reef.

Drifters

Microscopic plants, called plankton, drift in the sea. They give us much of the oxygen we breathe. Plant plankton is food for small drifting animals, which are in turn eaten by fishes, and even the blue whale, the largest animal on Earth.

Remember to put on the glasses to see plankton in 3D!

Mangrove Swamp

Mangrove swamps are found mostly in the tropics where fresh water mixes with salt water. The mangrove shrubs are partly submerged when the tide comes in. Prop roots anchor the plants in the mud.

British estuary

Estuaries are where rivers meet the sea. In cooler parts of the world, such as Britain, salt marshes are found in estuaries. Plants growing in the mud bind it together and help to protect the coast.

Rocky shore

When the tide goes out on a rocky shore, you can see different coloured seaweeds. Snorkel over the shore (in a safe place to swim) when the tide is in and the seaweeds look like a miniature forest.

Why is the sand wet?
Because the seaweed!

My snail cousins live at the seaside. Please colour in the seaweeds on their menu.

1.

2.

3.

1. Sea lettuce — I'm green. I live where fresh water runs down the shore.

2. Irish moss — I'm red. I live in rock pools and low down the shore. I am eaten in bramble flan in Ireland.

3. Kelp — I'm brown. I live low down the shore and in deeper water. Jelly taken from my leafy fronds keeps your ice cream soft.

9

Temperate House

You can find plants here from the warm temperate zone that is between the frozen poles and steamy tropics.

Chilli pepper

Olive tree

Date palm

Chilean wine palm

Coffin tree

Cycad
Kew's rarest plant

Balloon pea

Angel's trumpet

Toromiro
see Plants in danger

Cinchona tree

Knock, knock.
Who's there?
Leaf.
Leaf who?
Leaf me alone!

Did you know?
Temperate means not too hot and not too cold.

Put a sticker next to each plant on the list that you spot on your walk in the warm.

Find the plants

○ Angel's trumpet

○ Balloon pea

○ Chilean wine palm

○ Chilli pepper

○ Cinchona tree

○ Coffin tree

○ Cycad

○ Date palm

○ Olive tree

○ Tea bush

○ Thatch palm

You can get a butterfly's view of the plants if you climb the stairs.

Thatch palm

Tea bush

You can find out about these plants on the next few pages.

Inside the Temperate House

Discover how plants from warm temperate places keep us happy and healthy.

Time for a cuppa

Tea is made from the young leaves of the tea bush. Drinking tea started in Britain around 350 years ago. The Chinese have grown and drunk tea for over 2000 years.

Tea bush

Tall tales

Kew's wine palm stands over 17 metres tall. It is the world's tallest indoor plant. Palm wine is made in Chile from the sugary sap that drains out of a palm after it is cut down.

Chilean wine palm

Make a date

The date palm has 800 different uses. Here are just a few of them! The leaves are made into hats, rope and thatch. The trunk is made into shutters and doors. The dates are made into sweets and cakes.

Yemeni hat

Date palm

Dates

Mosquito

Life saver

Bark of the cinchona tree produces quinine, which is a bitter-tasting drug. This is still used to treat malaria. The deadly disease (passed on by mosquitoes) kills up to three million people a year.

Cinchona tree

Green olives turn black when left to ripen further.

Golden oil

Olives are fruit of the olive tree. The fruit and seed inside are crushed to release the oil, which is used in cooking. Olives are also eaten. But first they must be soaked in salty water.

Olive tree

out of Africa

Traditional remedies are important for healthcare in many parts of the world. In South Africa, the balloon pea is used as a tonic for people suffering from illnesses including AIDs and cancer.

Balloon pea or cancer bush

African hut

Join the dots to find out who is lurking in the pool.

Inside the Temperate House

Spot plants from warm places with some strange uses.
Keep a look out for Kew's rarest plant.

Chilli pepper

Red hot

Chillies are hot and spicy, which puts most animals off eating them. Many people like the taste. Birds cannot taste the fiery flavour. They eat the chillies and disperse their seeds when they poo.

Fresh chillies

Coffin tree

The coffin tree comes from Taiwan in Asia. The strong wood is made into coffins for grand funerals. Coffin trees grow as tall as 65 metres, taller than Kew's Pagoda.

Chinese coffin

Coffin tree

Thatch or kentia palm

Home and away

The thatch palm comes from Lord Howe Island off the east coast of Australia. It has been grown as a potted palm since Victorian times (mid 1800s). Most potted thatch palms you see have come from seedlings raised on this island.

Thatch or kentia palm

Angel's trumpet flowers

Angel's trumpet

Lonely boy

This male cycad grew from a stem collected from a solitary plant in South Africa over 100 years ago. No other plants were ever found. Today, this cycad only survives in gardens like Kew.

This cycad is Kew's rarest plant.

Buried alive

Angel's trumpet comes from Colombia in South America. In ancient times, the slaves and wives of dead Mayan kings were put to sleep by drinking a potion containing angel's trumpet, and then buried alive.

The male cycad has a cone but needs a female to make seeds.

Hello, I'm a sundew. Can you help me spot five different things in the Temperate House? The white box is a sticky trap like me that catches insect pests.

Answers at the back of the guide

15

Evolution House

Step back three-and-a-half billion years to begin your journey. Spot giant creepy crawlies along the way.

Liverworts

Early land plant
Cooksonia

Pillow-like structures found in Australia today.

Leap forward

Around 420 million years ago, there are small plants on land for the first time. Some of these become extinct, such as *Cooksonia*. Simple plants, called liverworts, are still with us today.

Clubmoss and giant dragonfly

Can you find the small mosses on the rocks and the giant millipede?

Tree ferns

Way back

Three-and-a-half billion years ago, the only plants on the planet are single cells. These plants live in the sea. They form mats of gooey green slime. Some mats are covered in layers of sand to form pillow-like structures.

Step into the swamp

Around 320 million years ago, giant clubmosses tower into the sky. Some are as tall as Kew's Pagoda. Ferns and horsetails grow much taller than they do today in the huge swamps.

Look for horsetails with their ribbed stems.

Did you know?
You can dig for fossils in Climbers and Creepers.

You can find butterflies in the gardens and in Climbers and Creepers.

Fossil fuel

As the giant clubmosses and other plants die, they sink into the swamp. Millions of years later, their remains have become fossilized and turned into coal.

Coal

You can find many conifer trees in the Gardens today.

Flowers flourish

Magnolia flower

Around 120 million years ago the first flowers bloom. They are pale and sweet-scented to attract beetles to carry their pollen. Showy colourful flowers only begin to bloom when there are bees, butterflies, and birds on the planet to pollinate them.

Jump into the Jurassic

Around 200 million years ago, conifers, cycads and other plants that produce seeds are doing well. Dinosaurs feast on ferns, horsetails and cycads.

Cycads have cones and make seeds.

Triceratops

Walk on to today

Walk in a dinosaur's footprints. Some dinosaurs eat the soft, juicy flowering plants. Flowers are all around us today, but dinosaurs go extinct 65 million years ago.

What kind of dinosaur slept all day?
The dino-snore.

17

Stag's horn fern

Titan arum

Tank plants

Agave

Water lettuce

Giant waterlily

Orchid

Axolotl
a kind of salamander

Barrel cactus

Aloe vera

Stone plant

Prickly pear

Gecko

Insects need to watch their step in this glasshouse. We are out on patrol ready to slurp up any pests.

Go in this way where the cacti are.

Pitcher plant

Sundew

Venus flytrap

Madagascan tree

Warm, wet or dry - which tropical climates do you prefer? You will find many climates in this amazing glasshouse.

Put a sticker next to each plant on the list that you spot on your walk around the desert and rainforest.

Find the plants

- Agave
- Aloe vera
- Barrel cactus
- Giant waterlily
- Madagascan tree
- Orchid
- Pitcher plant

- Prickly pear
- Stag's horn fern
- Stone plant
- Sundew
- Tank plants
- Titan arum
- Venus flytrap
- Water lettuce

You can find out about these plants on the next few pages.

Seek out plants that grow in deserts and other dry places.

Barrel cactus

Watery insides

The barrel cactus grows in American deserts, where it can survive for long periods of time on its store of water. The thick waxy outer layer keeps the water in and acts like sun-block.

Living stones

Stone plants look like stones, which may help them escape the notice of animals that might eat them. These plants grow in South Africa's drylands. When stone plants flower after the rains, they are much easier to see.

Stone plants in flower

Stone plants

Madagascan tree

Madagascan tree spines

Spiny trunk

This unusual tree comes from the dry forests of Madagascar. The spines on the trunk collect moisture from dew. It drips onto the ground then is sucked up by the tree's shallow roots.

20

Peel me first

The fruits of the prickly pear are covered in sharp spines. By carefully peeling the fruits, people can eat them. The orange flesh inside is also made into sweets.

Prickly pear

Did you know?
Conservatory is another name for a glasshouse.

Lookalike

Agaves look rather like aloes but they come from the other side of the world. Agaves live in dry places in Mexico, south-western USA and Central and South America. Each circle or rosette of leaves only flowers once, then dies.

Agave

Say aloe

Aloes come from Africa and Arabia. They survive in dry places by storing water in their thick leaves. The jelly-like sap from the leaves of aloe vera soothes the skin.

Aloe vera

Aloe vera lotion

Do you have shampoos or lotions made of aloe vera in your bathroom?

Cactus

Stone plant

Prickly pear

Agave

Aloe

Fit the plant names into the puzzle.

I

E

L

A

Answers at the back of the guide.

Inside the Princess of Wales Conservatory

Enter the lush wet zone where you may encounter giant leaves and flowers.
Explore more to see orchids and ferns.

Giant waterlily

Giant waterlily ribs

Water lettuce

Waterlily wonder

Giant waterlily leaves can grow over two metres across in summer. The leaves float because they trap pockets of air between the ribs underneath. The ribs are armed with spines to keep fish from nibbling at them.

The floating leaf can take the weight of a baby. Please do not try this yourself!

Lettuce menace

The water lettuce grows at the surface with its roots hanging down into the water. In some places, such as Florida in the USA, it is a menace. The dense growth blocks the light, and kills underwater plants.

The titan arum flower grows taller than a person.

Big stink

You may see the titan arum in bloom in summer. It is the biggest flower in the world and smells foul. The flower grows up from a huge corm (a type of bulb). The titan arum comes from the Sumatran rainforest in Indonesia.

A single giant leaf grows up afer the flower dies back.

Bromeliads on trees

Perfect pools

Tank plants perch high up on rainforest trees. The leaves collect water to make a little mid-air pool in which frogs sometimes raise their tadpoles.

Tank plant or bromeliad

Showy orchids

Orchids have beautiful, long-lasting flowers. Their seeds are tiny and blow away in the wind like dust. Most tropical orchids grow on tree branches, where their dangling roots catch moisture from the air.

Orchid on a tree

Tropical orchid

Fabulous ferns

The stag's horn fern grows on branches in the rainforests of south-east Asia. It is too high up to get nutrients from the soil. Instead, the fern collects debris in its leaves.

Stag's horn fern

Fern waterfall

Hello I'm an axolotl. Can you say my name 'ax-o-lot-al'? How many of my babies can you see in this pool? Come and see me in the Conservatory.

Answer at the back of the guide.

23

Inside the Princess of Wales Conservatory

Discover plants that eat animals, and go down the ramp to see who is underwater.

Have fun with us meat-eating monsters in Climbers and Creepers.

Sundew

Stickers

Sundews have droplets of glue to trap insects – just like flypaper. In some sundews, the leaf slowly wraps around the struggling victim to suffocate it. The dead insect is then digested.

Feed me

Carnivorous plants need to eat animals because they live in boggy places, where there are not enough nutrients in the soil. These meat-eating monsters have cunning tricks to trap small animals – most of which are insects.

Venus flytraps also like eating spiders.

Snap trap

An insect feeds on nectar at the base of the teeth but when it brushes against trigger hairs several times the leaf snaps shut. The teeth interlock to prevent the insect escaping. The leaf squeezes the insect and starts to digest it when still alive.

Slip and slide

An insect is lured deep inside a pitcher plant by large helpings of nectar. The deeper it goes, the more slippery the wall becomes. The insect slips and tumbles into the liquid at the bottom where it drowns and is digested.

Small animals, such as frogs, sometimes fall into large pitchers and get digested too.

Pitcher plant

Nasty nashers

Piranha fish

You can see a piranha in one of the smaller tanks downstairs. The piranha has powerful jaws armed with razor-sharp teeth that rip out chunks of flesh. Fortunately, most kinds are shy of people.

Giant catfish

Rainbow crab

Water dragon

Piranha   Discus fish

Stingray

Can you find the names of our animals in the word search?

Freshwater stingray

Sting in the Tail

If you look through the round tank windows downstairs, you can see freshwater stingrays. Their tails have sharp spines loaded with venom. Stingrays only attack when stepped on or harassed.

```
P I R A N H A X R U Q A S O V T W
C K O B Q B I P L E C U T X B M E
R A I N B O W C R A B N I B T S L
G I A N T C A T F I S H N Q F I V
Q L E K X T C X B T V O G U S E K
A Q D R M O W A T E R D R A G O N
K E X D N Y R T Z W Q B A F E X I
E J V K B X D T A F J E Y T H M D
G X D I S C U S F I S H E I D C N
```

Please note our animal displays change from time to time.   Answers at the back of the guide.

Plants and People Exhibition

Discover more useful plants and some beautiful and bizarre things made out of plants.

Cool hats

Hats made from dried leaves keep you cool in summer. The largest hat on display is made from leaves of the coconut palm. The panama hat shows how strips from the leaves of the hat plant are woven together.

Panama hat

Panama hat plant

Cannibal Cutlery

This fork and plate from the Fijian islands are made of wood from the intsia tree. They were given to Kew in the mid 1800s when fortunately the practice of eating people was coming to an end.

You would not be here without plants! We give you oxygen, food, shelter, medicines, clothes, and more.

Poisonous plants

People in the Amazon rainforest coat their arrows and blow-darts with different poisons. The poisonous roots of a vine are sometimes used for arrows. A drug made from this poison was also used to relax a patient's muscles during surgery.

Quiver for poison arrows

Heart pills

Chemicals from foxgloves have been used in heart medicines for over 200 years. You can find foxgloves growing in the Conservation Area in summer.

You can see a coconut palm and panama hat plant in the Palm House.

Ancient scroll

The stalks of papyrus were made into a material to write on in Ancient Egypt. The papyrus plant grows in wet places in tropical Africa including along the river Nile.

Papyrus scroll

Why did the tomato blush?

Because he saw the salad dressing.

Can you pick the correct answers to these questions about the Plants and People Exhibition?

1. What item of clothing on display is made from pineapple leaves?
a. socks **b.** shirt **c.** pants **d.** jumper

2. What plant produces a seed case, which you can use to scrub your back?
a. sponge **b.** bottle-brush plant **c.** prickly pear **d.** loofah

3. Half the world eats the seeds from which kind of grass?
a. rice **b.** corn **c.** bamboo **d.** rye

4. Wood from which tree is used to make cricket bats?
a. oak **b.** sycamore **c.** pine **d.** willow

5. What plant dyes your jeans blue?
a. blackberry **b.** indigo **c.** blueberry **d.** strawberry

6. What poisonous seeds are made into jewellery?
a. rosary **b.** deadly nightshade **c.** lupin
d. golden rain tree

7. What part of plants belonging to the cucumber family can you play music with?
a. leaves **b.** roots **c.** gourds **d.** flowers

Answers at the back of the guide.

Buildings old and new

Visit some beautiful old buildings.
Witness the latest in glasshouse design at Kew.

Keep us cool

Warm air is drawn out of the top of the Alpine House so that the mountain plants inside do not overheat. Cool air is blown onto the plants from an underground maze. This works in the same way as the cooling passages in a termites' nest.

The Davies Alpine House opened in 2006.

Kew Palace

This royal palace was once the home of George III (1760-1830). He spent time here as a child in the school room, and also when he fell ill with a disease that made him seem mad.

If you like paintings, why not visit the Marianne North Gallery to see flower paintings by this Victorian explorer.

Queen Charlotte's cottage

Did you know Kew was home to tigers as part of the Royal Family's collection of animals in the early 1700s?

Royal cottage

The thatched cottage was given to Queen Charlotte when she married George III. The Royal Family had picnics here. Kangaroos and exotic birds were once kept nearby.

Bombs away

This Chinese-style pagoda was built in 1762 for George III's mother. At that time it had 80 golden dragons perched on the roofs. In World War II bomb designers dropped dummy bombs from the top floor to test them.

The Pagoda

We are Japanese

Our Japanese Gateway is a much smaller copy of one in an ancient city in Japan. Can you see the gateway's beautifully carved flowers and animals? Stroll around the gateway to see what gardens look like in Japan.

Minka

Recycled house

The Japanese Minka is a house made of wood. It has a thatched roof and mud plastered walls. This traditional style house was taken apart in Japan and rebuilt here. Minkas withstand earthquakes, which are common in Japan.

Colour in this dragon from the Japanese Gateway. Can you find a dragon in front of the Palm House?

29

Terrific trees

Take a trek around the Gardens to track down some terrific trees.

Scramble up the Xstrata Treetop Walkway for an exciting view. Discover what lives around tree roots in the Rhizotron.

Xstrata Treetop Walkway

Climb 118 steps to walk amongst the treetops. Every season brings new views of our trees. High above the ground, you can see far into the distance.

This Japanese pagoda tree actually comes from China!

Prop me up

Nearly 250 years old, this pagoda tree is one of our oldest trees. Its lopsided look is probably because it grew with a double trunk - one half of which snapped away about 100 years ago.

Don't forget you can use the map of the Gardens to help you find these trees.

Unlucky tree

A small plane flew into the top of this tree in the early 1900s. Since then it has been struck by lightning twice. You can see the split in the bark left after it was struck in 1992.

Corsican pine planted 1814.

Big puzzle

The monkey puzzle tree comes from Chile and Argentina in South America. The tree got its name when first grown in England because a person joked it would be hard for a monkey to climb.

Monkey puzzle tree planted in 1846.

The oak's leaf has tooth-like edges like a sweet chestnut leaf.

Chestnut-leaved oak

What a whopper!

The chestnut-leaved oak is Kew's biggest tree. It was planted in 1846. The tree grows over 30 metres tall and the branches spread 30 metres across, which is nearly three times the length of a London bus.

How many trees grow at Kew?
a. 700 **b.** 5000 **c.** 8900 **d.** 14,000
Answer at back of the guide.

California giants

The coast redwood is the world's tallest tree towering over 115 metres. The giant redwood is the biggest with a trunk measuring 31 metres around. Redwoods take hundreds of years to reach record sizes. Our trees are youngsters with the oldest planted in 1912.

Giant redwood

Still living

Scientists thought the Wollemi pine became extinct two million years ago. In 1994 pines were discovered living in a gorge near Sydney, Australia. Kew is helping to grow more Wollemi pines because there are so few in the wild.

Wollemi pine planted in 2005.

Chinese tulip tree leaves

Save the trees

The American tulip tree was first planted at Kew in the 1770's. We also have 27 Chinese tulip trees that were planted in 2001. The Chinese tulip tree is rare because forests in China are being cut down.

American tulip tree

Spring at Kew

You can see millions of flowers, such as crocuses, daffodils, and bluebells.

Crocuses
February to March

Crocuses grow from bulbs, called corms, which are swollen stems whereas daffodil bulbs are swollen leaves. Over one-and-half million crocuses bloom in the Crocus Carpet near the Victoria Gate.

Glory of the snow
April

These flower after the snow melts in their mountain homes in Turkey, Crete and Cyprus. They are also called *Chionodoxa* - say *chee-ono-dox-a*.

Daffodils   February to March

Daffodils are bred for their colour, shape and scent. Wild daffodils grow in the woodlands and mountains of Europe including Britain. Like many spring bulbs, daffodil bulbs are poisonous, which stops animals eating them.

Bluebells
April to May

The British bluebell is at risk from cross-breeding with the Spanish bluebell, which has spread from gardens. You can spot a British bluebell because its flower stem droops to one side.

British bluebell

Wildlife watch

Visit the ponds and lakes to find these baby wild water birds.

Coot with young

Mallard ducklings

Spring gardening tips

If you want tall sunflowers in your garden or container in summer, you can give them a head start by planting seeds in small pots indoors in spring.

What do I carry from flower to flower so that they can make seeds? Fill in the puzzle to see the answer in the shaded squares.

1. When are tulips in bloom?
2. What kind of bulb is used with cheese to flavour crisps?
3. What part of a plant is green and traps sunlight?
4. What do bees do to get around quickly?
5. What chocolate treats do you eat in spring?
6. What is the name of the sweet substance in flowers?

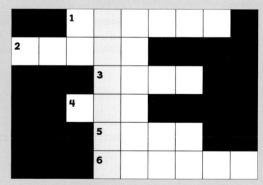

Answers at the back of the guide

Why do ducks watch the news?
To get the feather forecast.

Summer at Kew

On a warm summer's day why not bring a picnic and sit under the shade of the trees?

Roses

June to August

Just like bees, we are attracted to roses by their scent and colour. We can make perfumes from roses, and even a scented jam out of rose petals.

Bamboos

all year

You can see bamboos of different shapes and sizes in the Bamboo Garden. One of the Chinese bamboos in the Garden is the only food giant pandas eat.

Lavender

July to August

Lavender comes from the Mediterranean region across to India. They do not need much water so are good plants to grow when we have water shortages and a ban on using hosepipes in the garden.

Sunflowers

June to August

Look for sunflowers in the children's garden near Climbers and Creepers. If it is a good summer, they can grow taller than you.

Wildlife watch

Look for butterflies around flowers that are rich in nectar. Find damselflies and dragonflies around the lakes and ponds.

Emperor dragonfly

Painted lady butterfly

Blue-tailed damselfly

Can you help me back to my hive?

Summer gardening tips

Plant lettuce seeds in your garden or container outside. Keep the soil damp, and seedlings will appear about a week later. Soon you will be picking delicious lettuces.

What do insects learn at school?
Mathematics!

35

Autumn at Kew

The leaves of many of our trees change colour. You can see fruits, nuts, and mushrooms.

Japanese maple

The leaves of the Japanese maple in the Woodland Garden turn yellow with orangey edges. The bark of the young branches is a reddish orange.

Ginkgo

The female tree sprouts seeds that smell awful when they rot. If you pick up the seeds from the ground, their sickly smelling goo may irritate your skin. A stinky female tree is marked on the map.

Ginkgo leaves   Smelly ginkgo seeds

Sweet chestnut

Inside the green prickly fruit are two to three shiny chestnuts that are delicious when roasted. The ancient Romans were probably the first to grow sweet chestnut trees in Britain.

Sweet chestnut tree - one of the oldest trees in the garden.

Pampas grass

Look for pampas grass in the Grass Garden. This South American grass grows bigger than an adult person. It has large feathery seed heads.

Pampas grass

Why did the mushroom go to the party?

Because he was a fun guy to be with.

Wildlife watch

Spot squirrels and jays carrying off acorns to hide. Look for pine cones gnawed by squirrels to get at the seeds inside. Find fantastic fungi (mushrooms and toadstools) but remember some kinds are poisonous.

Jay

Squirrel

Poisonous mushroom

Autumn gardening tips

Pick up an acorn and plant it in a pot outside. Your baby oak should sprout in spring. When it out-grows the pot, you can put it in a larger container.

Can you work out which seeds belong to which tree?

Horse-chestnut

Ash

Oak

Pine

Answers at the back of the guide.

Winter at Kew

Wrap up warm and take a brisk walk around Kew's winter wonderland.

Holly

A good crop of holly berries is thought to mean a hard winter ahead. In fact, it means the summer weather was good so that many flowers set seed. We have over 56 different kinds of holly on Holly Walk.

Only female holly trees have berries.

Viburnum
October to March

You can smell the pink flowers from several metres away in the garden around the Ice House. In the 1700s ice cut from ponds was stored in the Ice House. The ice was used in summer to keep food and drinks cool.

Snowdrops
January

Snowdrops are some of the first flowers to bloom at the end of winter. Wild snowdrops grow in Britain across Europe to Turkey.

I spend most of the winter asleep in my sett.

Witch hazel
December to February

The curly yellowy-orange petals look like a burst of sunshine on a cold winter's day. Extracts from the bark and leaves are used to treat bumps and insect bites.

What does a caterpillar do on New Year's Day? Turns over a new leaf.

Wildlife Watch

Times are hard for birds in the winter months. They may have to find food under the snow.

Robin

Canada geese

Winter gardening tips

Wear gloves if you have sensitive skin. Balance a hyacinth bulb on top of a narrow jam jar full of water. The roots will grow to fill the jar. The bud opens into a scented flower.

Who made these tracks in the snow? Match the animal with its tracks.

Robin

Duck

Squirrel

Fox

Answers at the back of the guide.

39

How does your garden grow?

Kew has more than 100 people who work hard to keep the plants you see around you in good shape.

Big on bulbs

One of the biggest jobs in autumn is to plant more bulbs. Did you know it took 12 people one week to plant 750,000 bulbs at Kew even with a digger?

Never ends

The planting, pruning and weeding never stops. There are jobs to do all through the year.

I go out looking for food when everyone has gone home to bed.

Smelly heaps

We make our own compost from dead and waste plants, and truckloads of horse manure from the army and police stables. On cold days, the gigantic piles of compost steam because they get so warm from the rotting plant matter and dung.

Wildlife friendly

We like to encourage wildlife at Kew. You can watch birds in the hide in the Conservation Area and join in a pond dipping session at the Dipping Pond.

Meet the tree gang

The tree gang are expert at climbing trees. They use ropes and pulleys like mountain climbers. They have to saw off branches that might otherwise fall on your head.

Stag beetle sculpture in the Loggery

Stag beetle heaven

The Loggery is made of rotting logs, home to the grubs of Britain's largest beetle, the stag beetle. The grubs spend over four years feasting on the rotting wood before turning into adults.

You can see a giant tropical millipede in Climbers and Creepers.

How many of my mini-beast friends are living in the log pile?

Woodlouse   Millipede   Centipede

Ground beetle   Earwig   Stag beetle grub

Answers at the back of the guide.

Kew does not use peat because digging for peat destroys the boggy places where sundews like me grow.

41

Plant hunters

For over 200 years people from Kew have travelled the world in search of plants.

Beginning with Banks

Sir Joseph Banks (1743-1820) was a friend of George III and an intrepid explorer. Banks sent the first plant hunters from Kew on expeditions to bring back rare and useful plants.

Sir Joseph Banks

In the Himalayas

Hunting for new kinds of rhododendrons in the Himalayas was gruelling for Sir Joseph Dalton Hooker (1817-1911), who became head of Kew. He was infested with 100 blood-sucking leeches after tramping the foothills.

Sir Joseph Dalton Hooker

Rhododendrons

Violent volcano

When a volcano on the Caribbean island of Montserrat erupted in 1997, much of the island's forest was destroyed. Scientists from Kew helped the islanders preserve the remaining forest.

The volcano is still active.

Jungle adventures

Today, Kew helps people around the world to look after their plants. Expeditions to the rainforest in Cameroon, West Africa aim to discover which plants there are under threat.

Wild food

Kew's scientists are looking at traditional foods in Africa, such as roasted seeds from the sausage tree. Traditional foods may help people in Africa stay healthy and find enough to eat in times of famine.

The sausage tree is named for the shape of its fruit – not because it grows sausages!

Marvellous medicines

Many people around the world rely on plant medicines. Scientists from Kew study plants to find new medicines.

What vegetable can you use to fly a kite?
String beans.

Which plant lives where?
Match the plant to where it lives

Harebell

Desert

Rocky shore

Stag's horn fern

Cactus

Grassy meadow

Seaweed

Rainforest

Answers at the back of the guide.

Plants in danger

More and more plants are at risk as forests are chopped down, and wild places are taken over by towns and farms.

Time for trees

Much of the dry forest on the coast of Peru in South America has been cut down for farming and fuel wood. We are working with local people to plant more huarango (say wa-ran-go) trees.

Children help to plant trees in Peru.

Our world may be getting warmer because of people's energy-guzzling lifestyles. Plants like me may die out.

Wild no more

The last wild toromiro tree was chopped down for firewood in 1960 on Easter Island in the Pacific Ocean. New trees have grown from seeds planted at Kew.

You can see a toromiro tree in the Temperate House.

Bottle palm

This palm only grows wild on Round Island, Mauritius in the Indian Ocean. In the 1970s, it almost died out because rabbits and goats like to eat it. Seedlings from Kew have been planted back in the wild.

You can see the bottle palm in the Palm House.

Some of our rarest plants are grown in the laboratory.

Back from the dead

A schoolboy discovered the café marron plant in 1980 on the Indian Ocean island of Rodrigues. People thought it had been extinct for 40 years. We grew cuttings from this plant, and sent the young plants back to the island.

You can see café marron in the Palm House.

Did you know?
Kew grows 15 kinds of plants that are extinct in the wild.

The seeds I collected have got in a muddle. Can you help me find 13 peanuts in this jar of seeds?

Peanut

Seed bank saviours

About 700 million seeds from around the world are kept safe at Kew's Millennium Seed Bank at Wakehurst Place. We are collecting and storing seeds from places where plants are at risk, and we are helping other countries to set up their own seed banks.

Answers Page

Pages 4 and 5
Number of ants on the pages = 32.

Page 7 Word scramble
Make a shelter using **palm** leaves and **bamboo** poles.
Take cane from the **rattan** palm to make a chair.
Collect some **coconuts, bananas, sugar cane** and **mangoes** to eat.

Page 13 Who's lurking in the pool?
Koi carp

Page 15 Spot the difference
mushrooms, pool of water, leaf beside bucket, leaves in bucket, leaf on tree

Page 21 Word puzzle

Page 23
Number of baby axolotls in pool = 5
The other 8 were adults!

Page 25 Word search

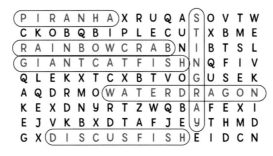

Page 27 Plants and people quiz
1b. shirt, 2d. loofah, 3a. rice, 4d. willow, 5b. indigo, 6a. rosary, 7c. gourds

Page 31 Number of trees
d. 14,000

Page 33 Word puzzle

Page 37 Matching game

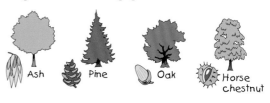

Ash   Pine   Oak   Horse chestnut

Page 39 Matching game

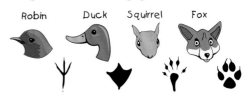

Robin   Duck   Squirrel   Fox

Page 41 Numbers of mini-beasts in the log pile
Woodlouse = 4
Millipede = 3
Centipede = 3
Ground beetle = 4
Earwig = 3
Stag beetle grub = 3

Page 43 Matching game
Harebell lives in **grassy meadow**.
Stag's horn fern lives in **rainforest**.
Cactus lives in **desert**.
Seaweed lives on **rocky shore**.

My name is Sticker the sundew.

46

Here are some things to do at home

I'm a harebell — I live in wild grassy places. You can help wildflowers like me and other wildlife by doing some of these things.

Dig a pond.
Pile up old logs.

Build a bird table and a bird bath.
Put up a bird or bat box.

Grow wild flowers to help bumblebees.

In dry summers use dirty bathwater to water plants.

Leave grass and nettles to grow in a sunny place in your garden as food for caterpillars.

Use a water-butt to collect rainwater for watering your plants.

47

PLANTS PEOPLE
POSSIBILITIES

First published in 2007, reprinted with amendments in 2008 and 2010.
by the Royal Botanic Gardens, Kew,
Richmond, Surrey, TW9 3AB, UK

Written by Dr Miranda MacQuitty
Illustrated by Guy Allen
Designed by Louise Millar

Great care has been taken to maintain the accuracy of the information
contained in this work. However, neither the publisher, the editors nor
authors can be held responsible for any consequences arising from use of
the information contained herein.

ISBN 978 1 84246 163 1

www.kew.org

Printed in England by Beacon Press

The following people have provided invaluable help and advice:
Raffat Bari, Shirley Beadle, Sandra Bell, Fiona Bradley, Gail Bromley, Karen Brown,
Henry and Rose de Chazal, Colin Clubbe, Lucy Cole, Simon Cole, Louise Cross,
Ambar and Hannah Driscoll, Gina Fullerlove, Laura Geary, Christina Harrison,
Tina Houlton, Tony Kirkham, Lloyd Kirton, Jane, Emily and Michael Lambert,
Paul Little, John Lonsdale, James and Oliver Morley, Christine Newton,
Jill Preston, Sue Seddon, Varsha, Shivani and Ria Sokal, Katie Steel,
Nigel Taylor, Catherine Welsby, Lydia White.

We would like to thank the following for providing photographs and for
permission to reproduce copyright material:
Acro Images/Alamy 25; Heather Angel/Natural Visions 35; Jeff Collett/
Natural Visions 8; Emma Dodd 37; Redmond Durrell/Alamy 25; Jeff Eden 40;
Peter Gasson 23, 33, 35, 36, 37; Hemera Technologies/Alamy 27;
Images&Stories/Alamy 35; Laura Jennings 23; Christabel King 43;
Douglas Lander/Alamy 7; Paul Little cover, 1, 5, 8, 12, 13, 17, 26, 32, 43;
William Milliken 43, James Morley cover, 22; Norman T Nicoll/Natural Visions 8;
Brian Rogers/Natural Visions 16; Pep Roig/Alamy 21; Brian Schrire 34;
StockStill/Alamy 37; Scott Taylor 34; Oliver Whaley 44; all other photographs
by Andrew McRobb.

For information or to purchase all Kew titles please visit
www.kewbooks.com or email publishing@kew.org

All proceeds go to support Kew's work
in saving the world's plants for life

The Royal Botanic Gardens, Kew, is a world famous scientific organisation,
internationally respected for its outstanding collection of living plants,
world-class herbarium, and scientific expertise in plant diversity,
conservation and sustainable development. Kew is also a major visitor
attraction comprising 132 hectares of historical landscape.

Kew's mission is to inspire and deliver science-based plant conservation
worldwide, enhancing the quality of life.

When you find a plant at Kew put a sticker on the plant lists in the guide.

Daffodil

Rubber tree

Stag's horn fern

Corsican pine

Titan arum

Cycad

Snowdrop

Tulip tree

Chilli pepper

Glory of the snow

Rattan palm

Bamboo

Giant waterlily

Pampas grass

Lavender

Monkey puzzle

Bluebell

Rose

Balloon pea

Thatch palm

Sweet chestnut

Tea bush

Giant bamboo

Pitcher plant

Cinchona tree

Stone plant

Aloe vera

Cycad

Gingko

Witch hazel

Agave

Coffin tree

Venus flytrap

Japanese maple

Viburnum

Japanese
pagoda tree

Annatto
tree

Giant
redwood

Can you find the stickers for
the trees and other plants
you spot in the Gardens?

Coconut
palm

Tank
plants

Coco-de-mer

Sunflower

Barrel
cactus

Vanilla
orchid

Date
palm

Holly

Madagascan
tree

Chestnut-
leaved oak

Angel's
trumpet

Orchid

Crocus

Sundew

Cacao
tree

Sugar
cane

Prickly
pear

Chilean
wine palm

Wollemi
pine

Coffee
bush

Water
lettuce

Can you find the stickers
for plants you spot
in the glasshouses?

Olive
tree